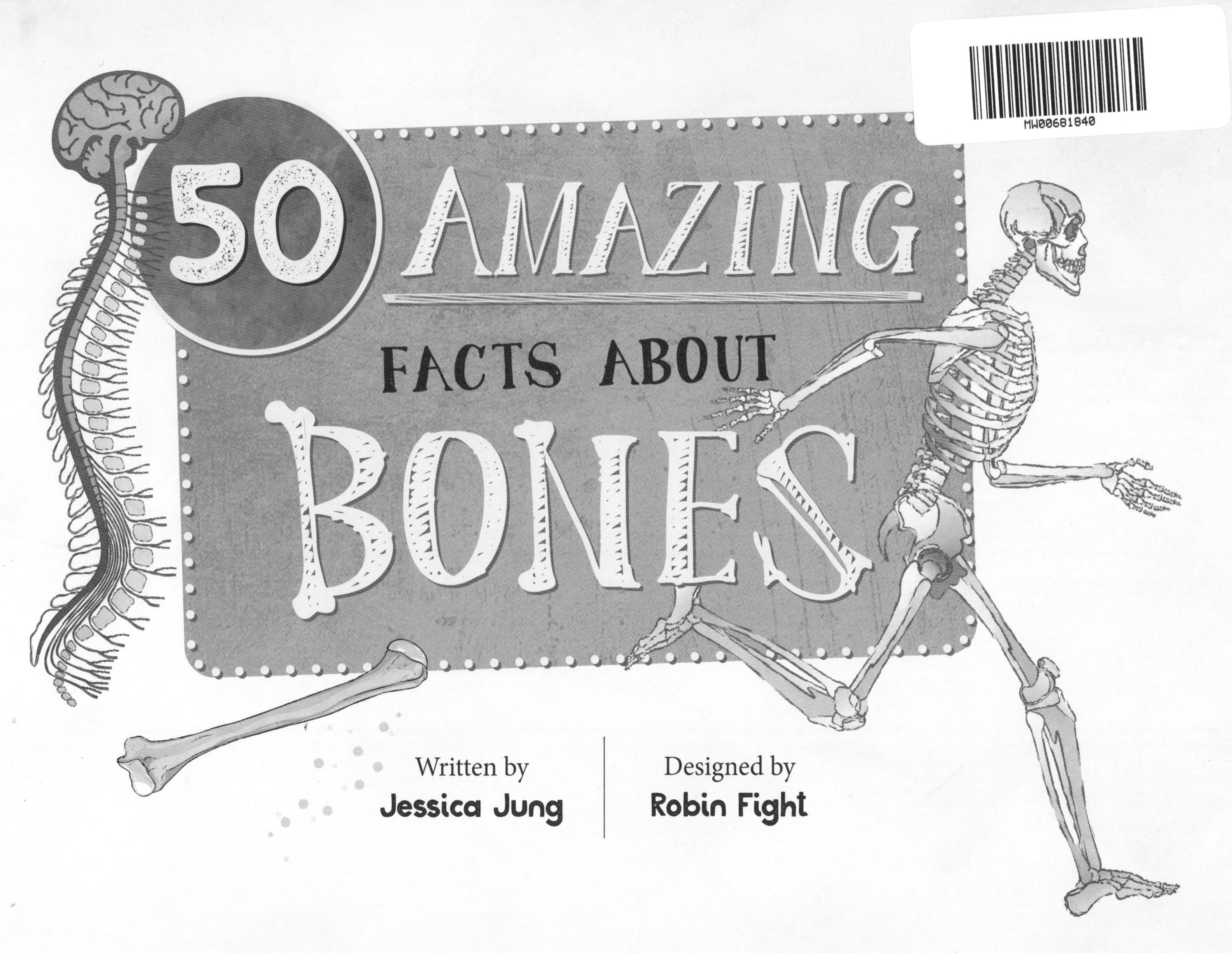

50 AMAZING FACTS ABOUT BONES

Written by
Jessica Jung

Designed by
Robin Fight

From the tops of our heads to the bottoms of our feet, our bodies are held up by amazing hard tissues called bones. There is so much to learn about bones and how they impact living things, so let's take a closer look!

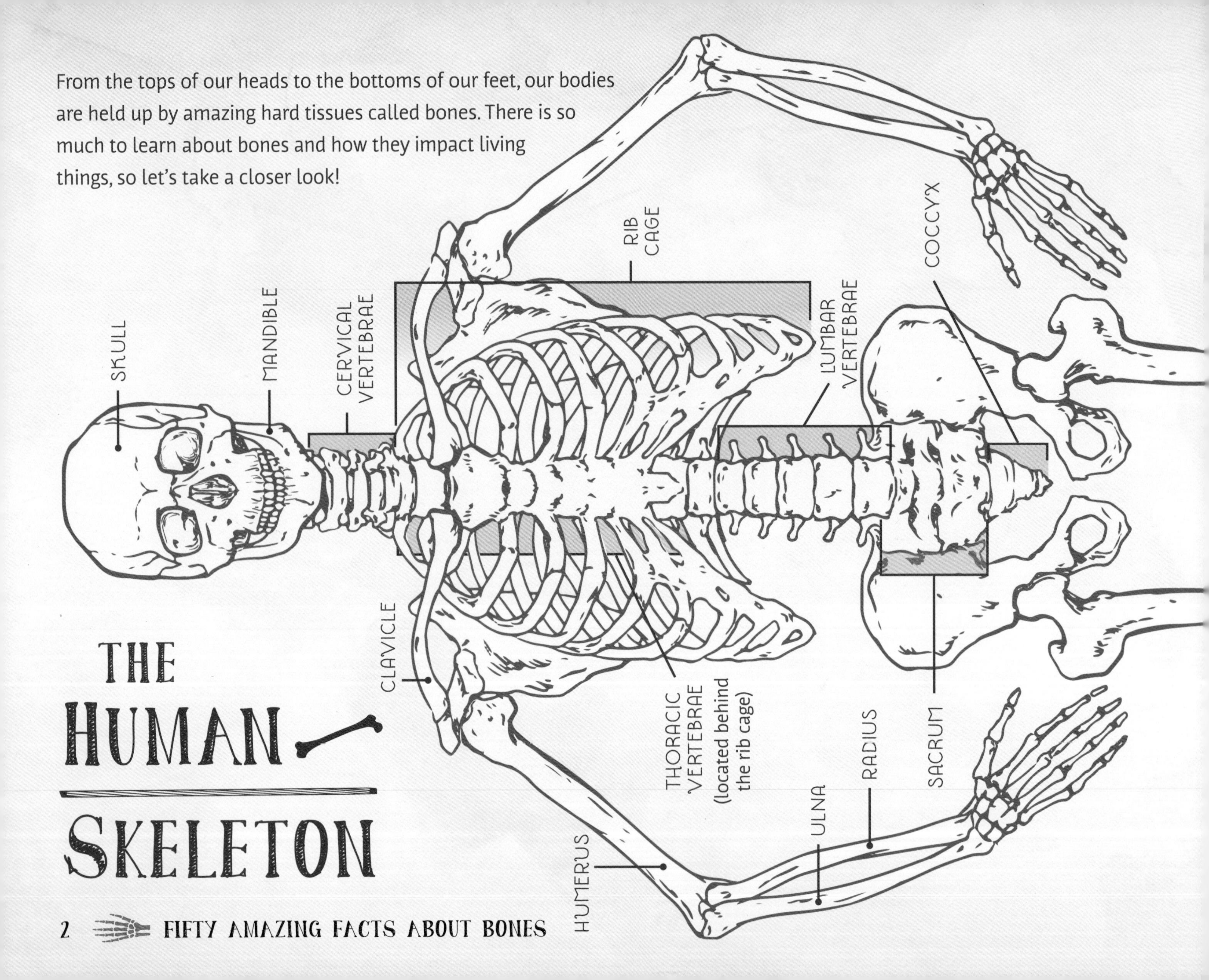

THE HUMAN SKELETON

DID YOU KNOW?

All mammals have bones—not just humans! Birds, amphibians, reptiles, and most fish also have bony skeletons. None of these animals could survive without bones.

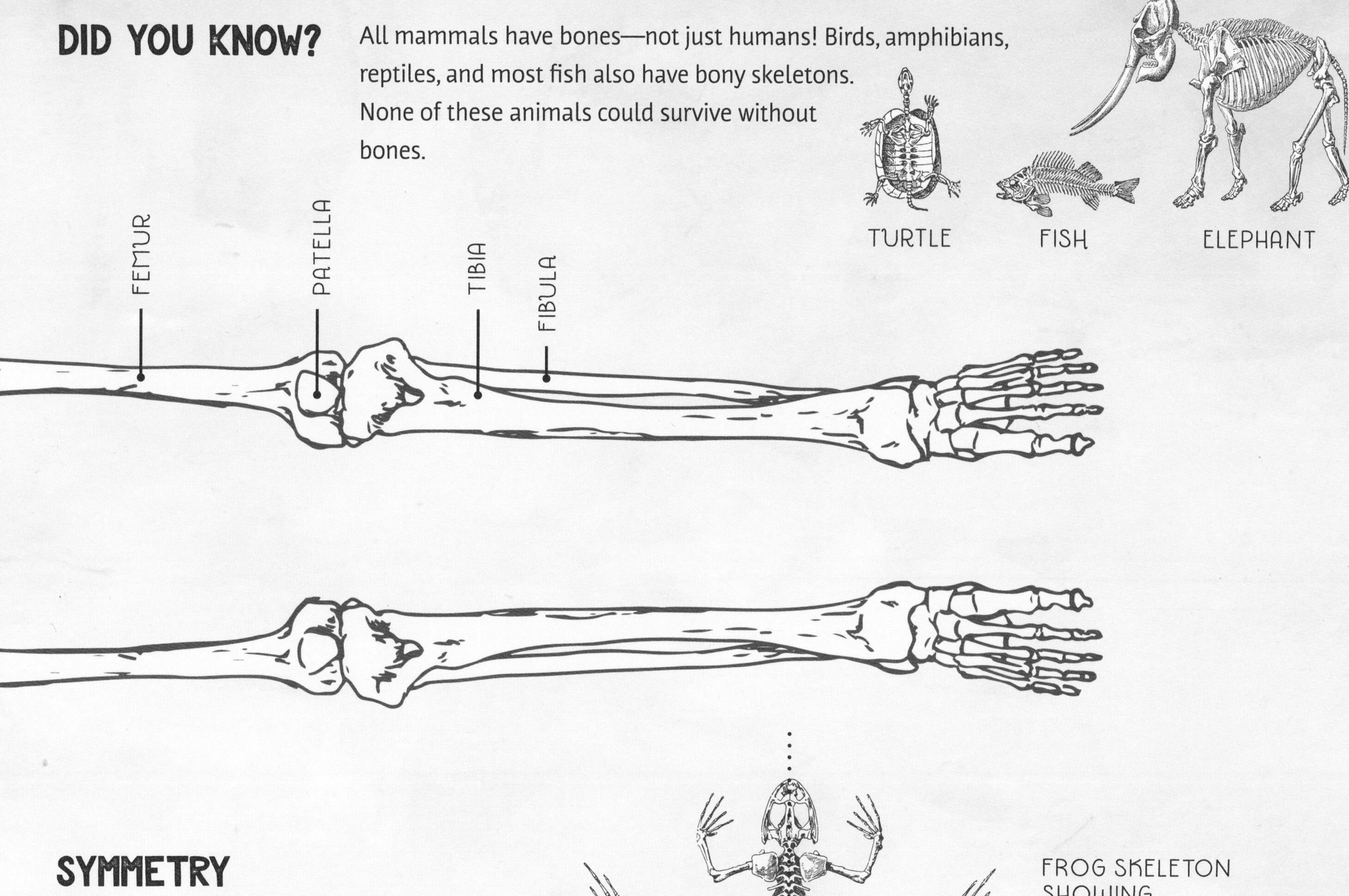

SYMMETRY

Humans and most animals have bones that are symmetrical on both sides of the body.

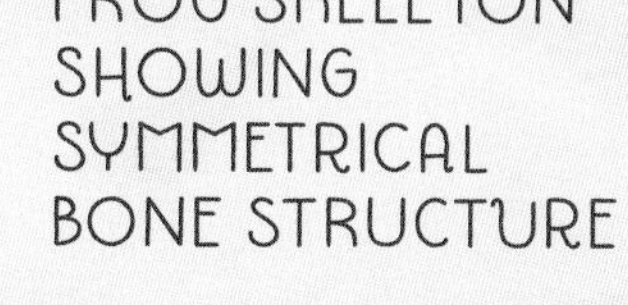

FROG SKELETON SHOWING SYMMETRICAL BONE STRUCTURE

Bones are alive! They grow, strengthen, and—amazingly—can even repair themselves!

The human ***skeletal system*** is made up of bones that provide

1. a framework of support,
2. protection for our delicate internal organs, and
3. the ability to move our bodies.

SKELETAL SYSTEM

FACT 03

Without bones you would be a squishy, shapeless blob of skin and organs, resting helplessly on the floor, unable to move or walk about.

GROWTH

FACT 04

Did you know human babies have more bones than adults? It's true! Babies are born with about 300 bones.

A baby has about

300

bones.

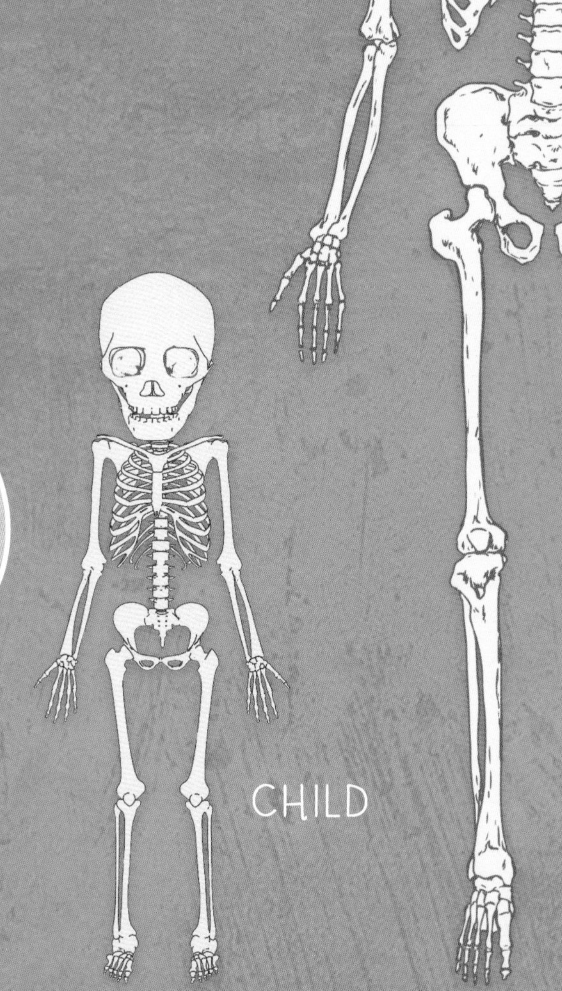

FACT 05

As you grow, some of your bones begin to fuse, or grow together, and some of your ***cartilage*** (a flexible, rubbery connective tissue) turns to hard bone, eventually reducing your total amount of bones to 206.

A full-grown adult has

206

bones.

FACT 06

Your kneecaps were not always bone! Babies are born with kneecaps made of cartilage. Between the ages of 2 and 6, this cartilage starts to solidify into bone through a process called ***ossification***.

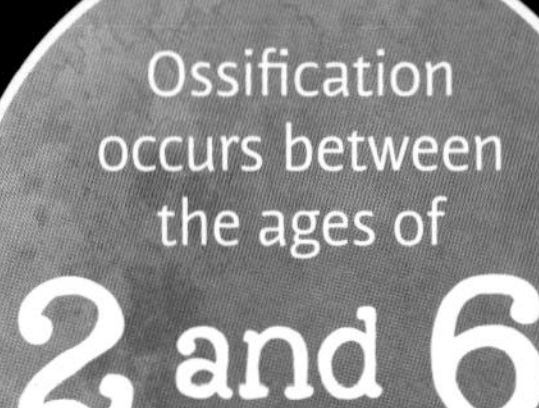

X-RAY OF A BABY'S LEGS

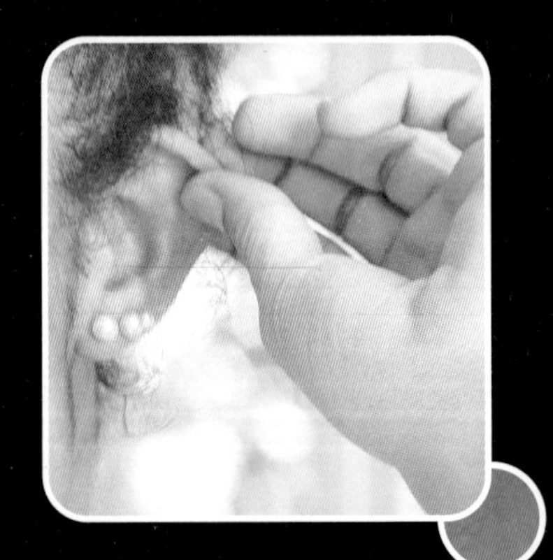

FLEXIBLE CARTILAGE OF THE EARS AND NOSE

FACT 07

Not all cartilage gets replaced by hard bone as you get older. Gently try to bend your outer ears and wiggle the tip of your nose. These two body parts will remain cartilage your whole life. In fact, cartilage is found in several other places in the body, such as those cushiony discs between the bones in your spine, the bronchial tubes in your lungs and airways, the ends of your ribs, and even at the bendable joints between bones.

FACT 08

Did you wake up taller today? It's possible! Your bones do most of their growing at night while you're sleeping.

X-RAY OF A CHILD'S KNEES

EPIPHYSEAL PLATES

FACT 09

The location where new bone growth forms is called a growth plate, or ***epiphyseal*** [ep-uh-FIZ-ee-uhl] ***plate***. It is made of cartilage and is found toward the ends of the long bones in your arms, legs, hands, and feet. This growth plate will permanently "close" once you've reached your full height, usually around the age of 15 for girls and 16 for boys. How much more growth do you still have?

Made of cartilage, your bones' epiphyseal plates will close once you have grown to your full height.

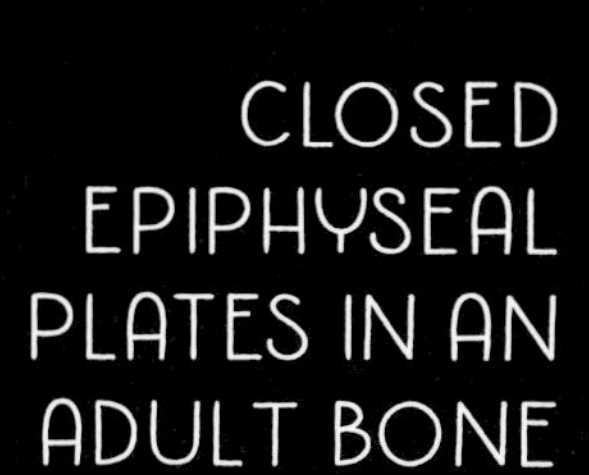

CLOSED EPIPHYSEAL PLATES IN AN ADULT BONE

FACT 10

Every 10 years, our bones fully replace themselves in a process called ***remodeling***. This removal of old bone cells and the replacement of new cells happens without your even knowing or feeling it!

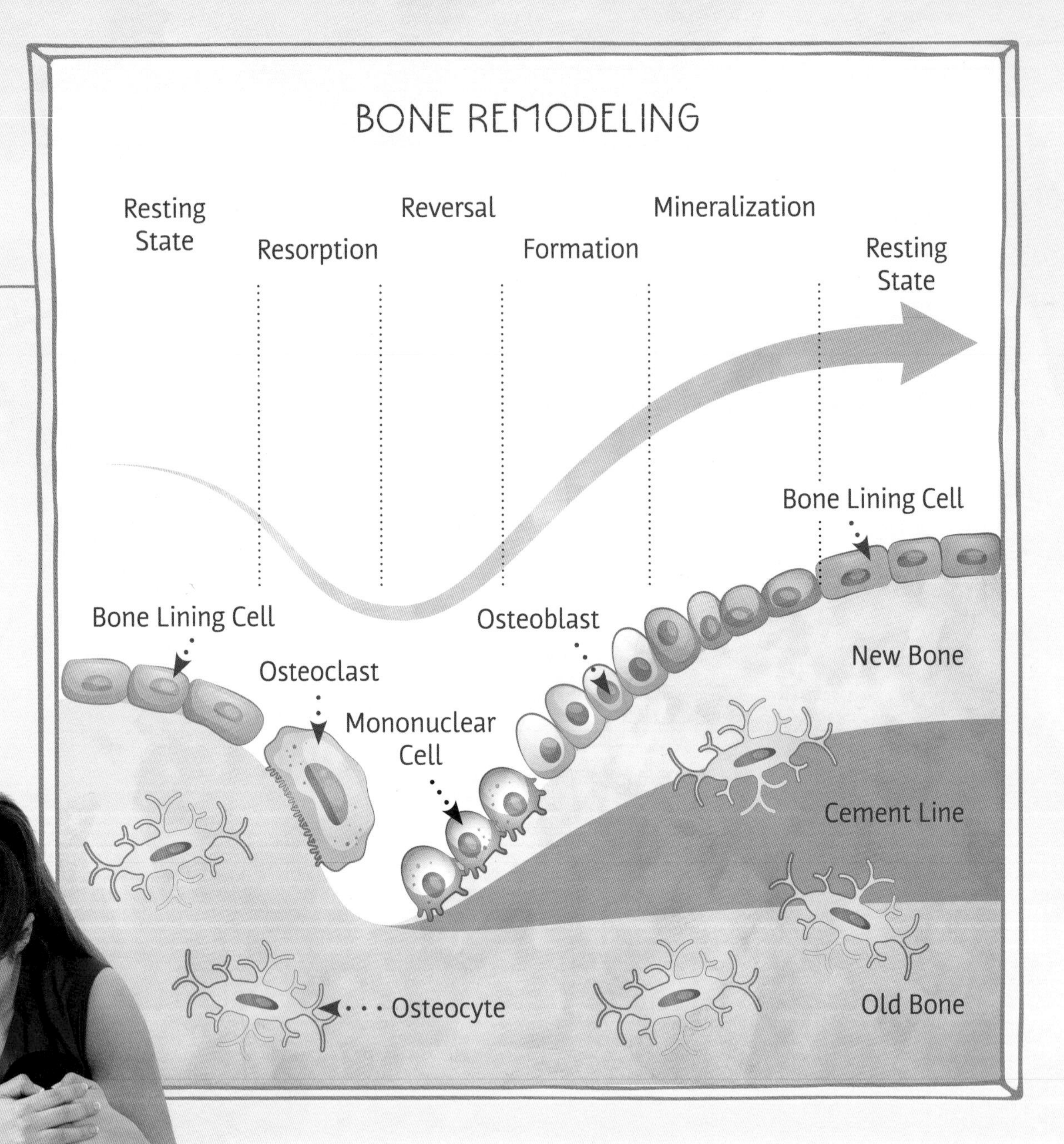

FACT 11

Contrary to popular belief, "growing pains" are not actually from growing bones but rather aches and pains originating in the muscles. Thankfully, normal bone growth is painless!

FACT 12

Your bones will stop growing at some point, but they don't stop changing! Around the age of 40, bones start to lose density, or mass. Calcium, vitamin D, and weight-bearing exercise, such as walking or running, can help strengthen bones and prevent them from thinning and getting brittle.

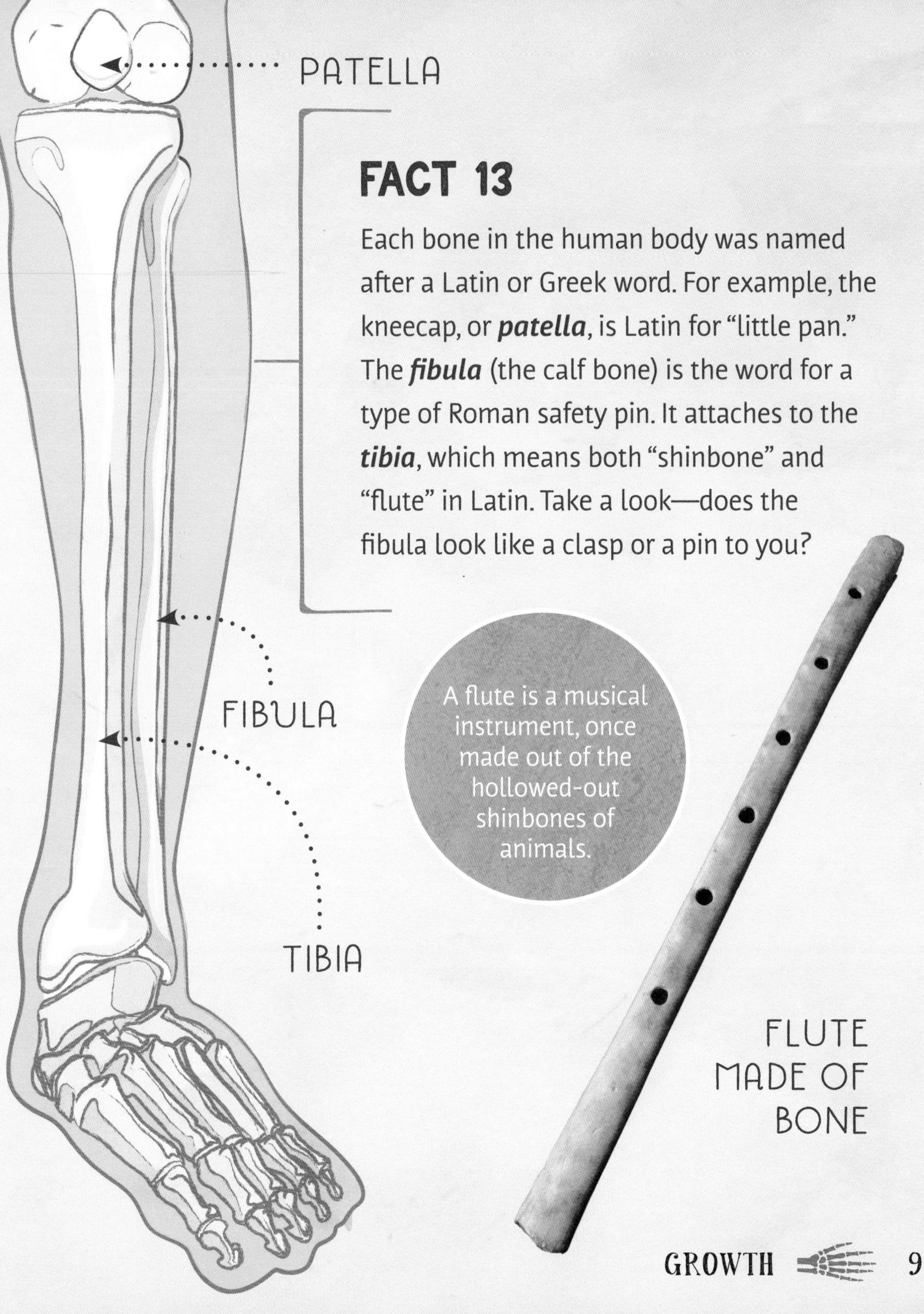

FACT 13

Each bone in the human body was named after a Latin or Greek word. For example, the kneecap, or ***patella***, is Latin for "little pan." The ***fibula*** (the calf bone) is the word for a type of Roman safety pin. It attaches to the ***tibia***, which means both "shinbone" and "flute" in Latin. Take a look—does the fibula look like a clasp or a pin to you?

A flute is a musical instrument, once made out of the hollowed-out shinbones of animals.

STRUCTURE

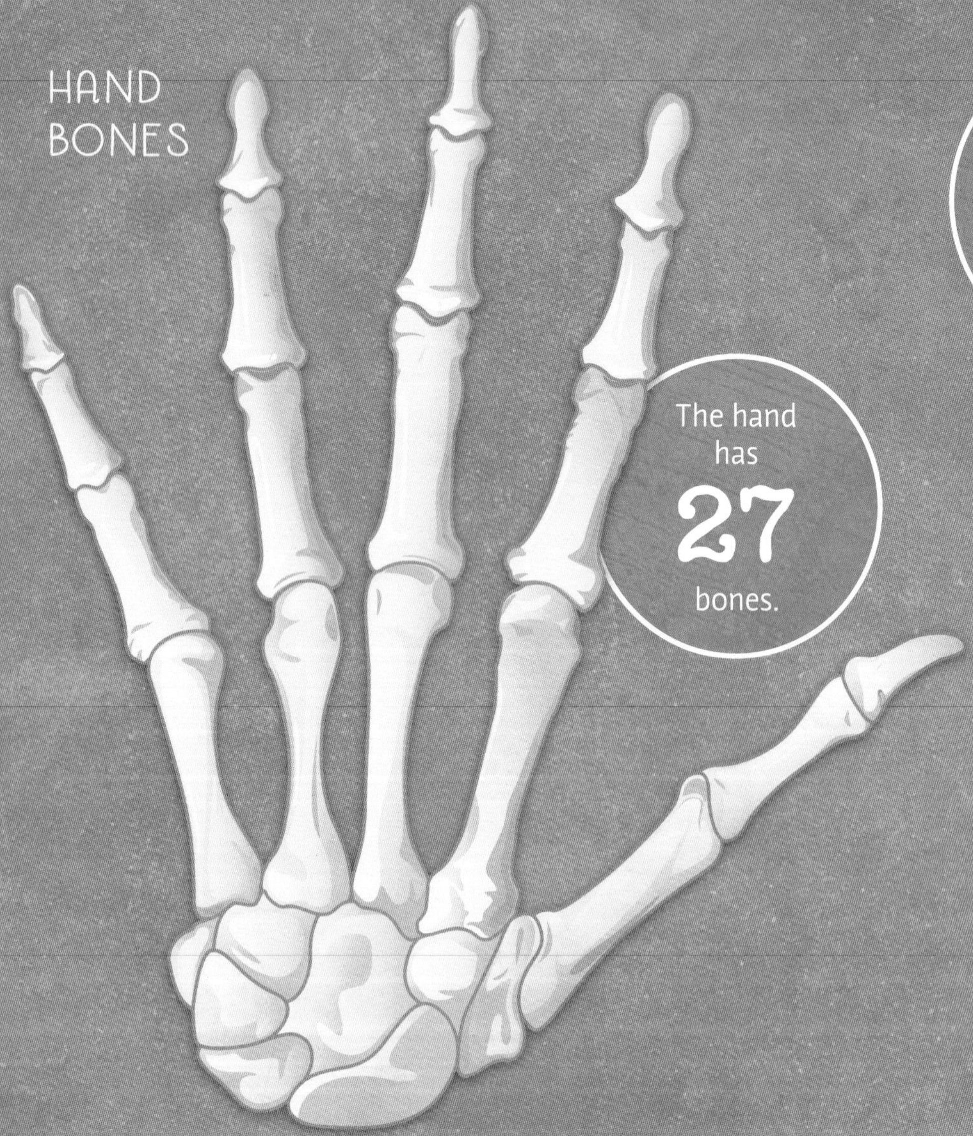

HAND BONES

The hand has **27** bones.

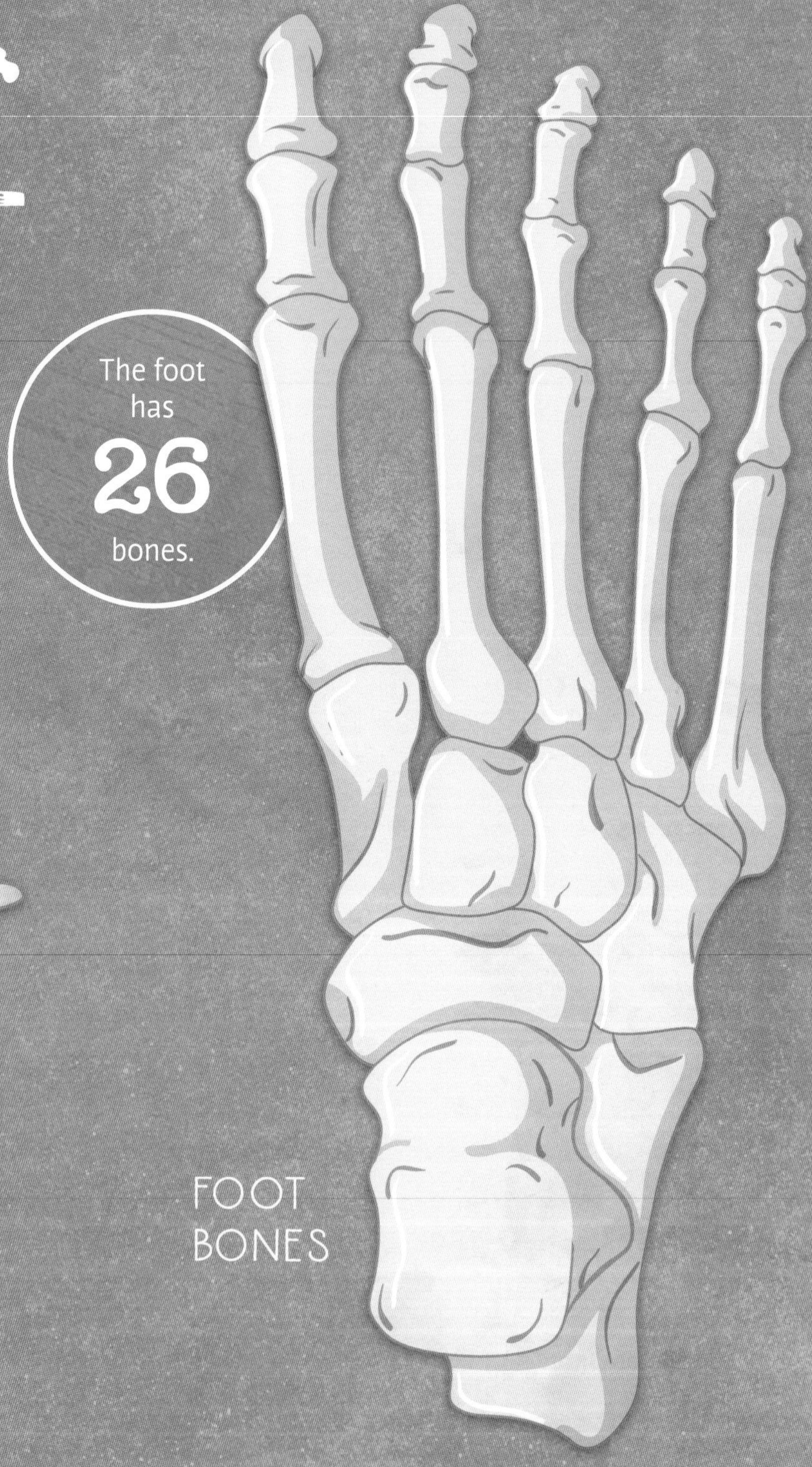

The foot has **26** bones.

FOOT BONES

FACT 14

Each bone fits perfectly in the exact place it was designed for.

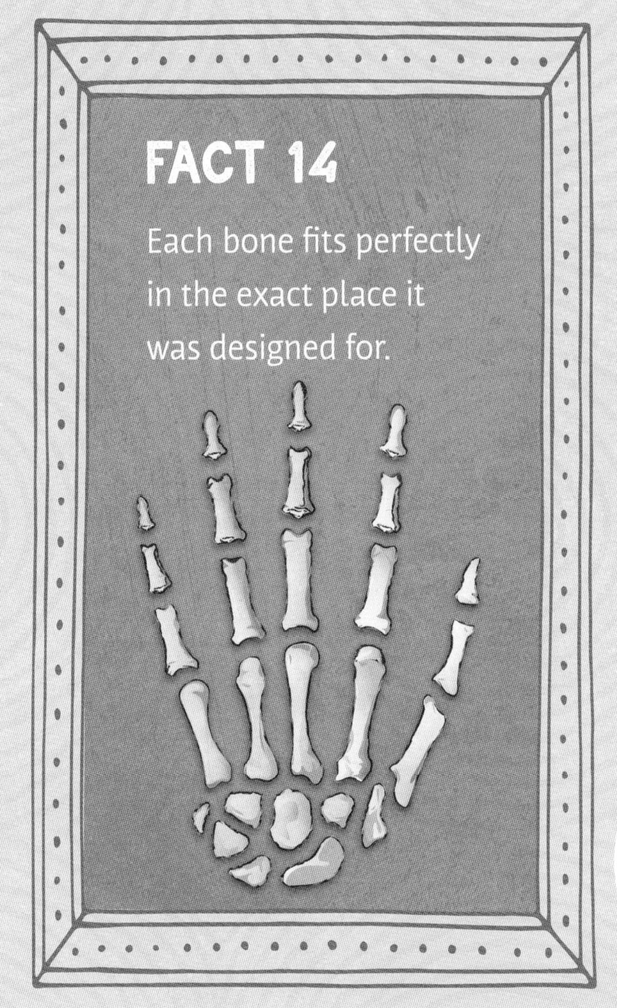

FACT 15

Each hand has 27 bones, while each foot contains 26 bones. Altogether, both hands and feet total 106 bones—making up more than half the bones in the human body!

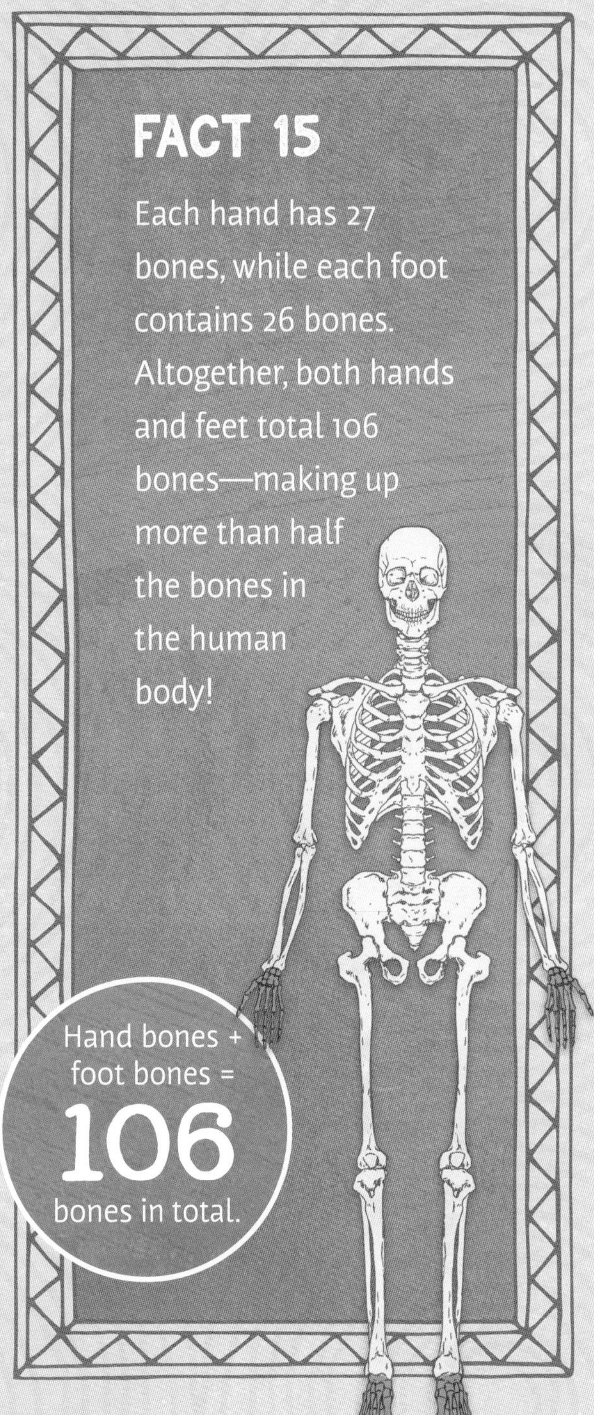

FACT 16

The "funny bone" is not actually a bone at all, nor is it very funny when it gets bumped! That tingling sensation you sometimes get when you accidentally hit your elbow is really the result of your ulnar nerve getting knocked against your ***humerus*** [HYOO-muh-ruhs], the long arm bone between your shoulder and elbow.

HUMERUS

FACT 17

Your ***spine***, that long, bumpy column of bones supporting your back, is made up of 33 individual bones called ***vertebrae***, and each one is shaped like a ring. This extremely important column of bones has the hefty job of encasing your ***spinal cord***—a large bundle of nerves that sends signals from your brain to the rest of your body.

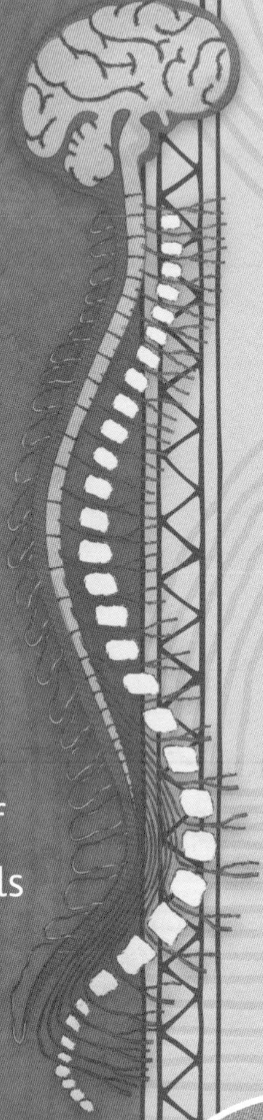

SPINAL CORD

The spine has **33** vertebrae.

FACT 18

Five different regions make up the spine: ***cervical***, ***thoracic***, ***lumbar***, ***sacrum***, and ***coccyx***. The upper 24 vertebrae allow your back the flexibility to bend, arch, and twist. The lower nine vertebrae are more tightly connected with less range of motion. The lowest section (the coccyx) is commonly known as the tailbone. It helps provide balance and stability when you are sitting.

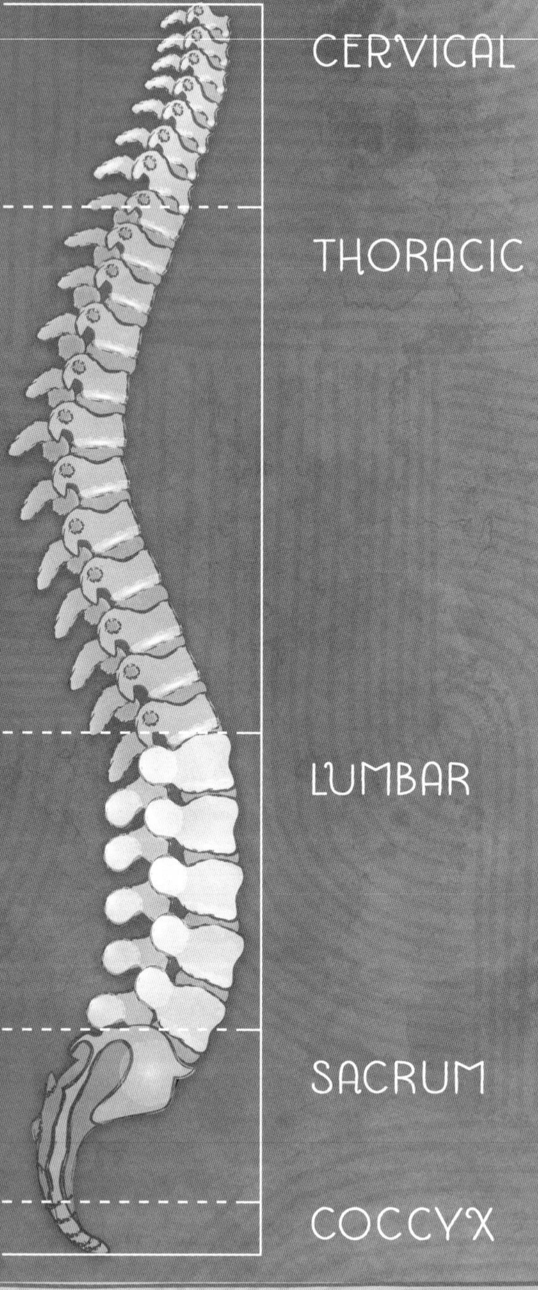

CERVICAL

THORACIC

LUMBAR

SACRUM

COCCYX

FACT 19

The spine is extremely strong! It has over 120 muscles and 220 ligaments supporting it. The spine of an average healthy adult can support the weight of hundreds of kilograms of pressure!

FACT 20

Giraffes and humans both have seven neck (or cervical) bones—except the bones in a giraffe's neck are about 15 times longer than ours!

Giraffes and humans have **7** neck bones.

FACT 21

Did you know that a baby's spine is the first set of bones to grow when the baby is being formed inside the womb?

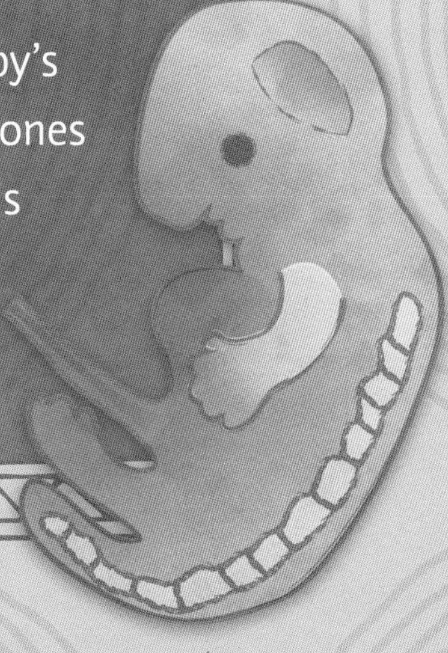

BABY FORMING IN THE WOMB (NOTICE THE SPINAL COLUMN)

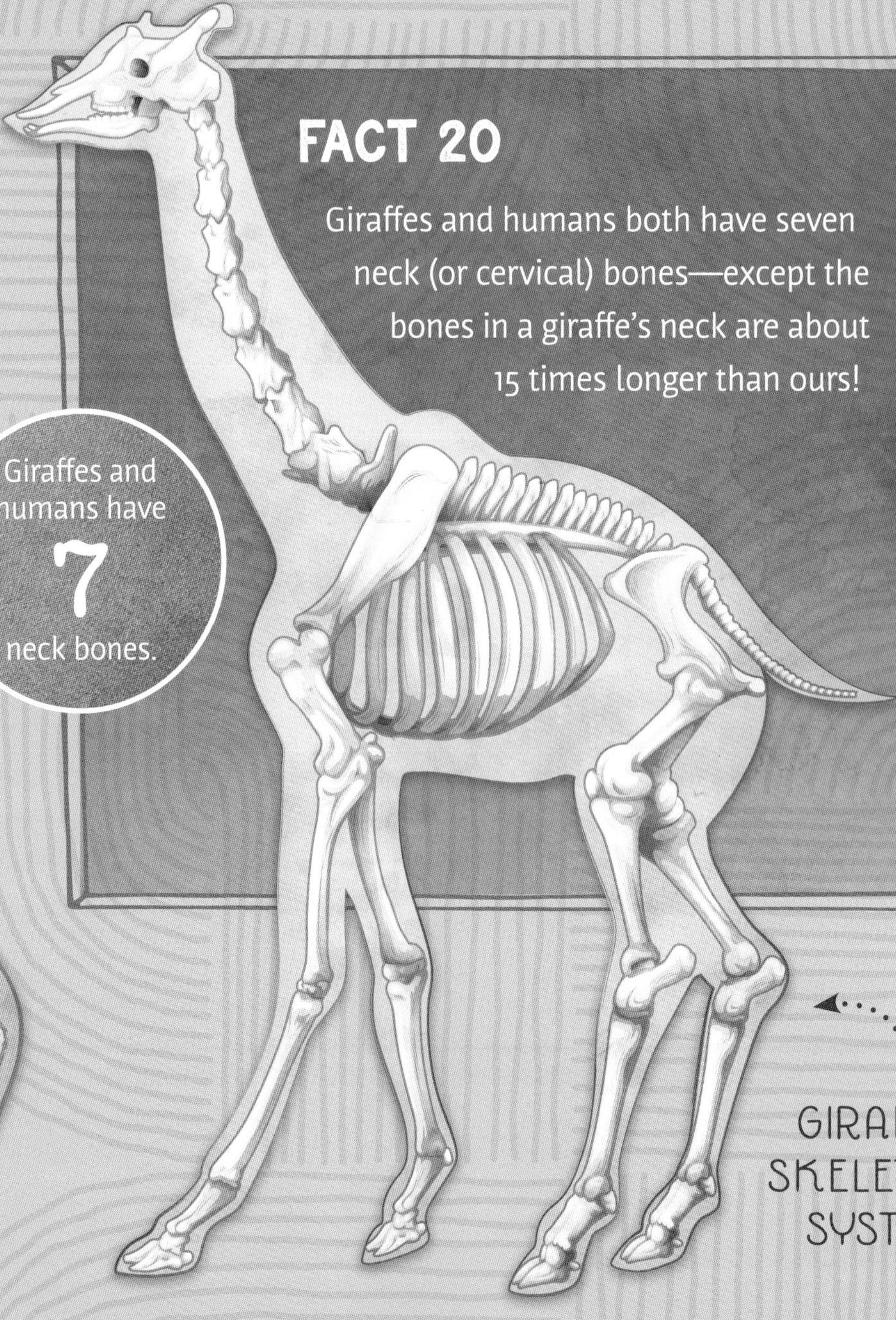

GIRAFFE SKELETAL SYSTEM

SIZE, WEIGHT & SHAPE

FACT 22

Your entire skeletal system accounts for about 14% of your total body weight.

TOTAL BODY WEIGHT

14% Skeletal System

FACT 23

The longest and strongest bone in the human body is the thighbone, known as the ***femur***. It measures about one-fourth of your body's total length.

The femur is almost $\frac{1}{4}$ of the body's length.

FEMUR

FACT 24

Have you ever visited or seen pictures of the Eiffel Tower in Paris, France? The tower's curved iron beams were inspired by the shape of the curved upper portion of the femur and its natural strength in holding up weight. Furthermore, the femur's sturdy crisscross-shaped bone fibers also inspired the metalwork pattern of this iconic tower.

FACT 25

Your smallest bone is the ***stapes*** [STA-pees]. This stirrup-shaped tiny bone in your middle ear is responsible for sending sound vibrations to your inner ear. Although minuscule in size, measuring only 2–3 mm (0.08–0.12 in), this bone plays a big role in proper hearing function.

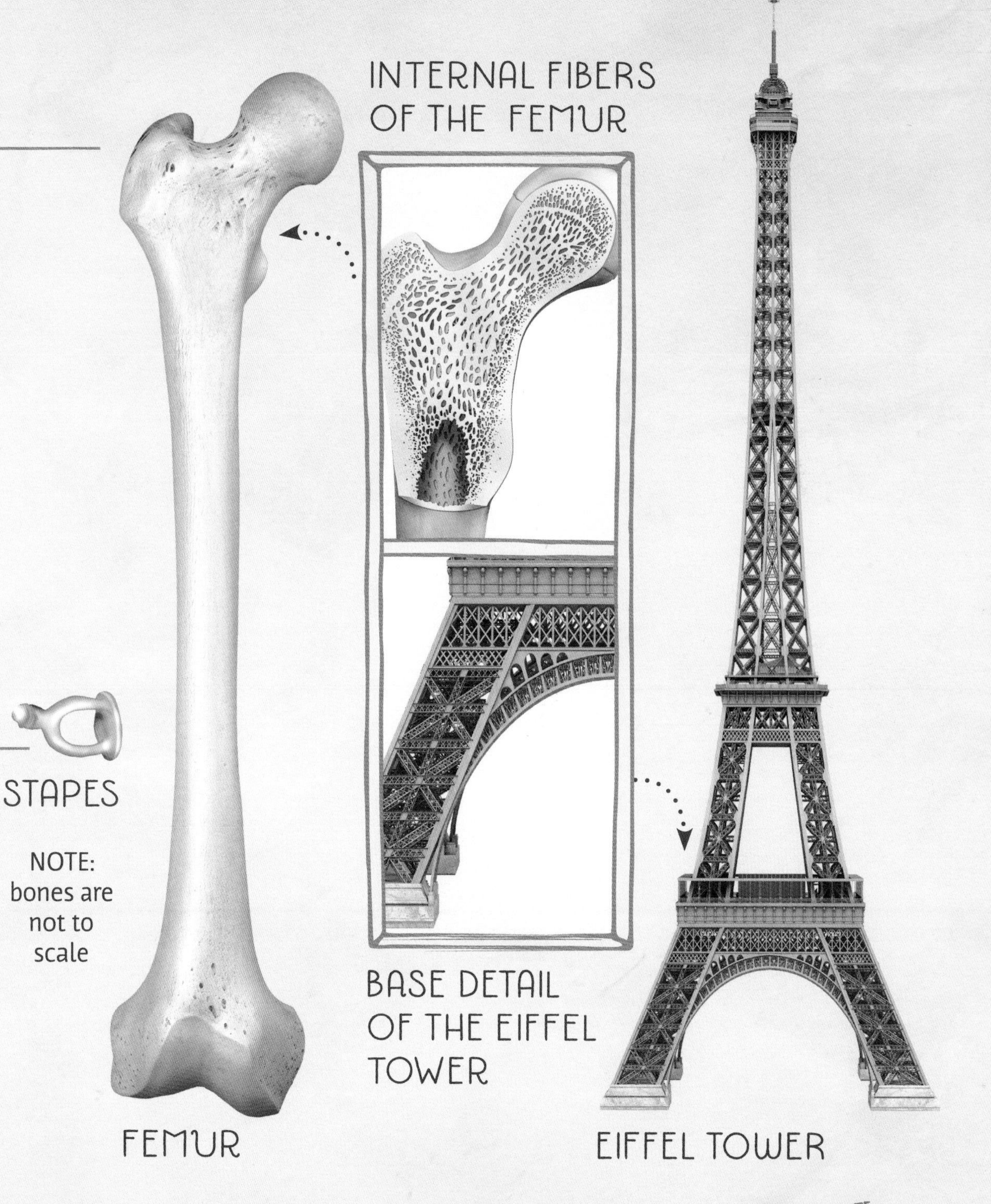

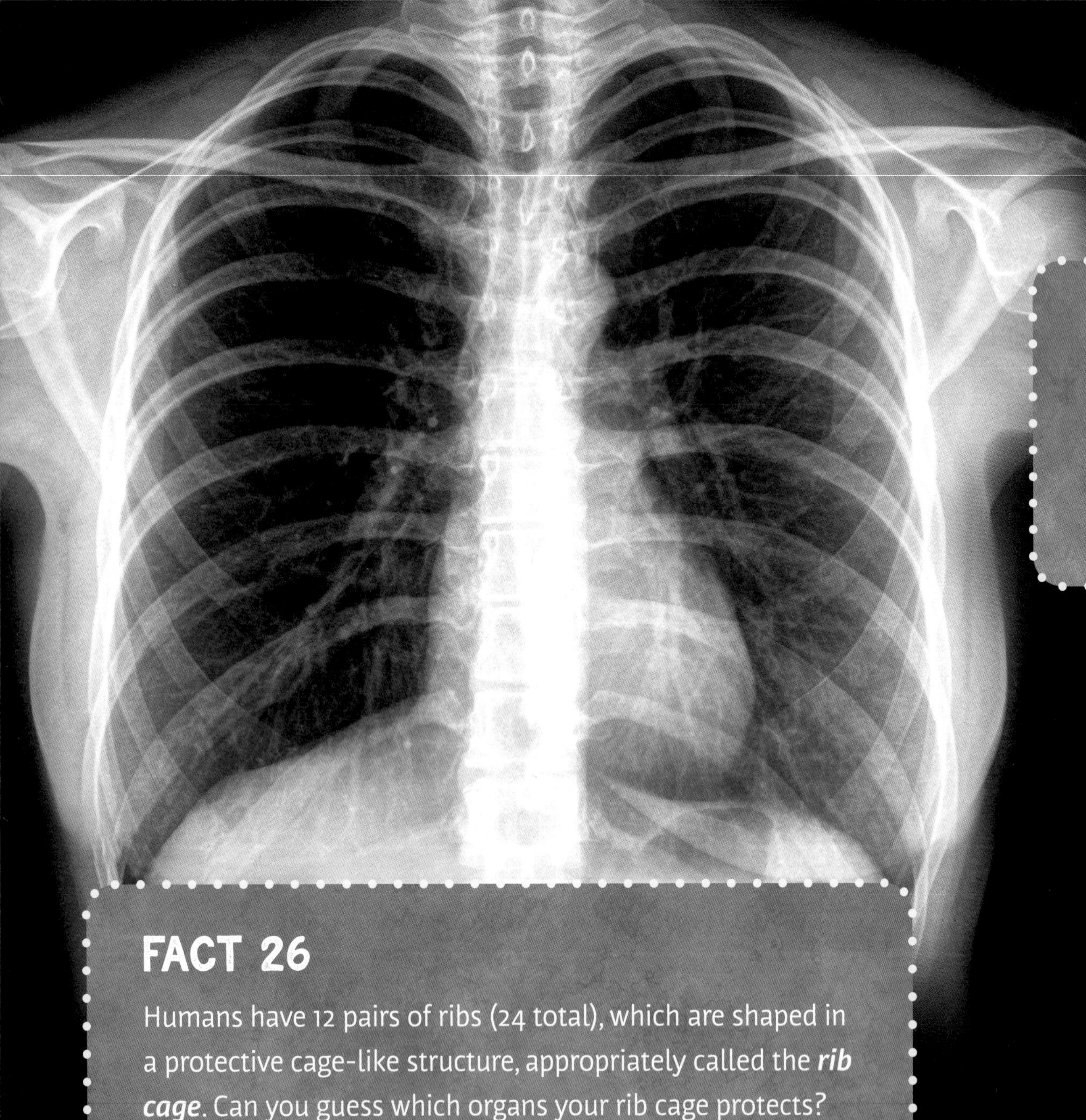

X-RAY OF AN ADULT RIB CAGE

FACT 27

One out of every 200 people is born with an extra rib, called a ***cervical rib***, located at the base of the neck, above the ***clavicle*** (collarbone).

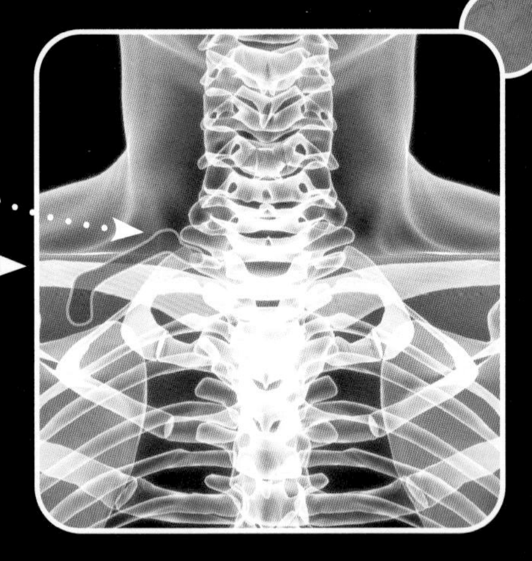

CERVICAL RIB (shown in red)

CLAVICLE

FACT 26

Humans have 12 pairs of ribs (24 total), which are shaped in a protective cage-like structure, appropriately called the ***rib cage***. Can you guess which organs your rib cage protects?

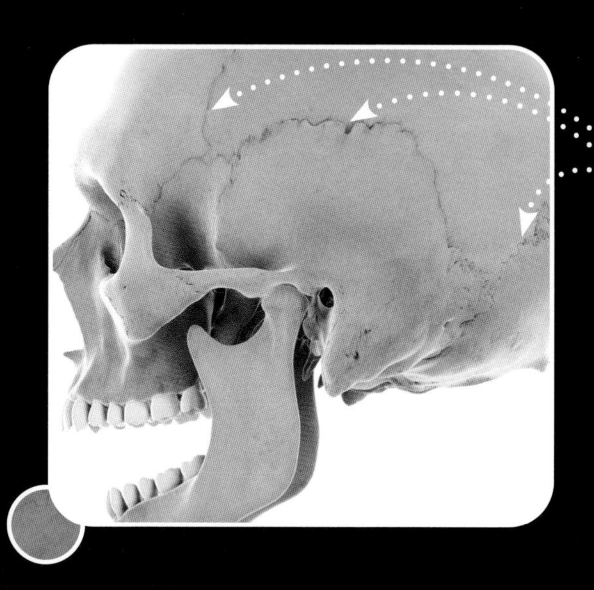

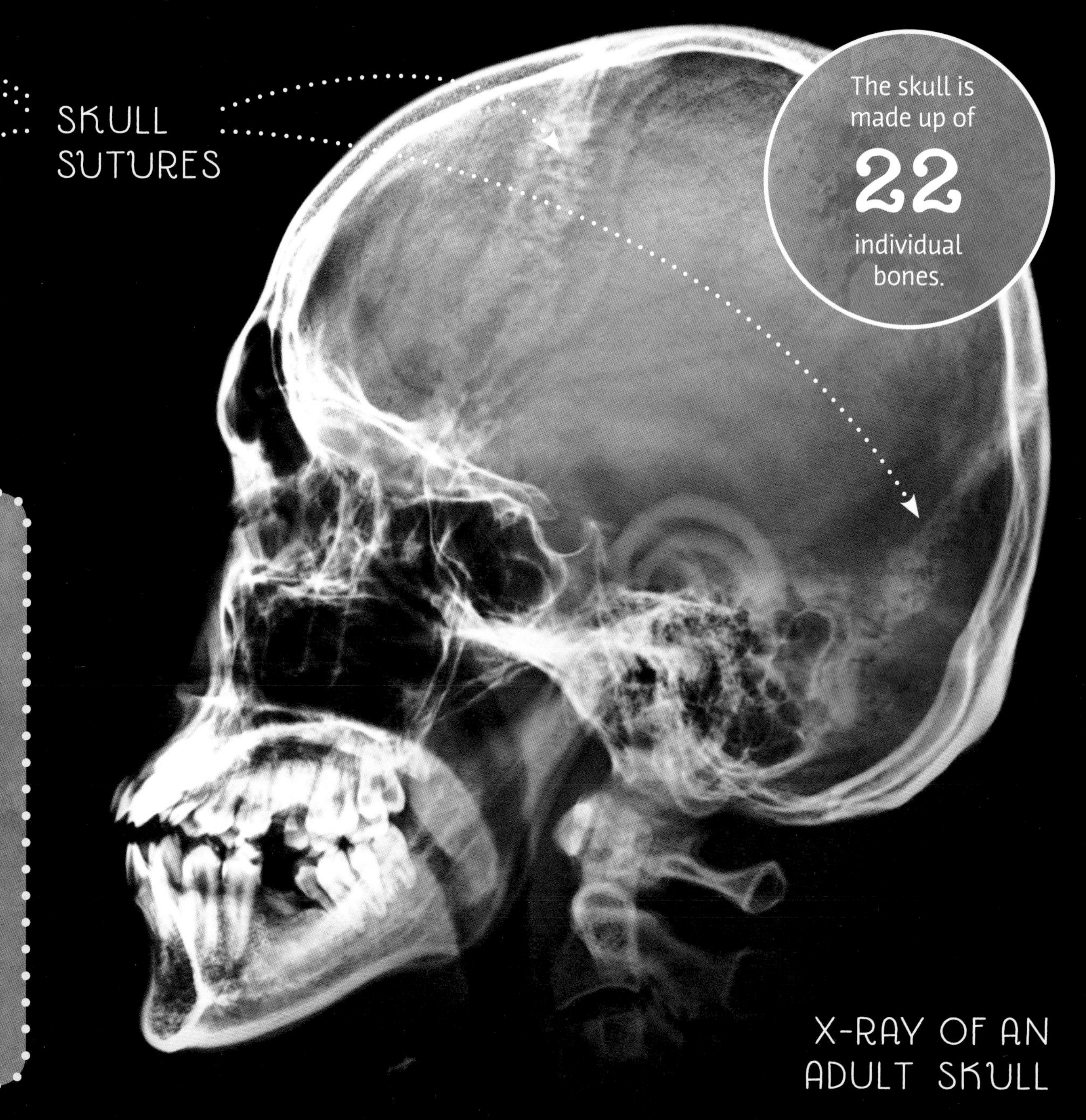

X-RAY OF AN ADULT SKULL

FACT 28

While it may seem like your skull is one round piece of bone, it's actually made up of 22 individual bones designed to protect your delicate brain and give your face its unique shape. These bones are all held together by strong, fibrous tissue at fixed (or immovable) joints called ***sutures***.

MOVEMENT

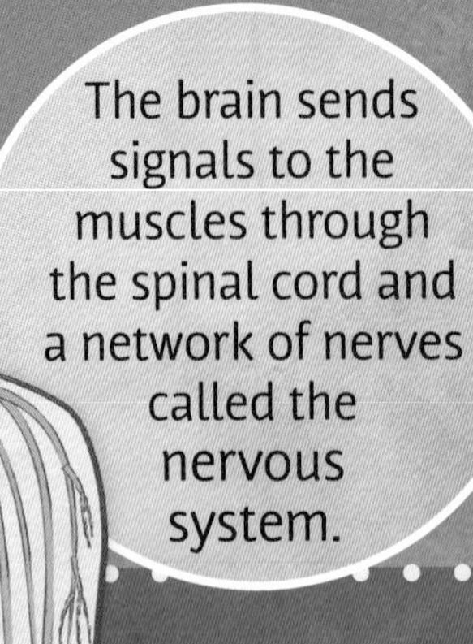

The brain sends signals to the muscles through the spinal cord and a network of nerves called the nervous system.

FACT 29

Bones don't move on their own. That's right! It's your brain that must send signals to your muscles, which are attached to your bones by ***tendons***. The tendons then pull on your bones to make them move.

FACT 30

Bones are attached to other bones by straps of long, stretchy, fibrous tissues called ***ligaments***. Think of a stretchy rubber band!

NERVOUS SYSTEM

LIGAMENTS OF THE FOOT AND ANKLE

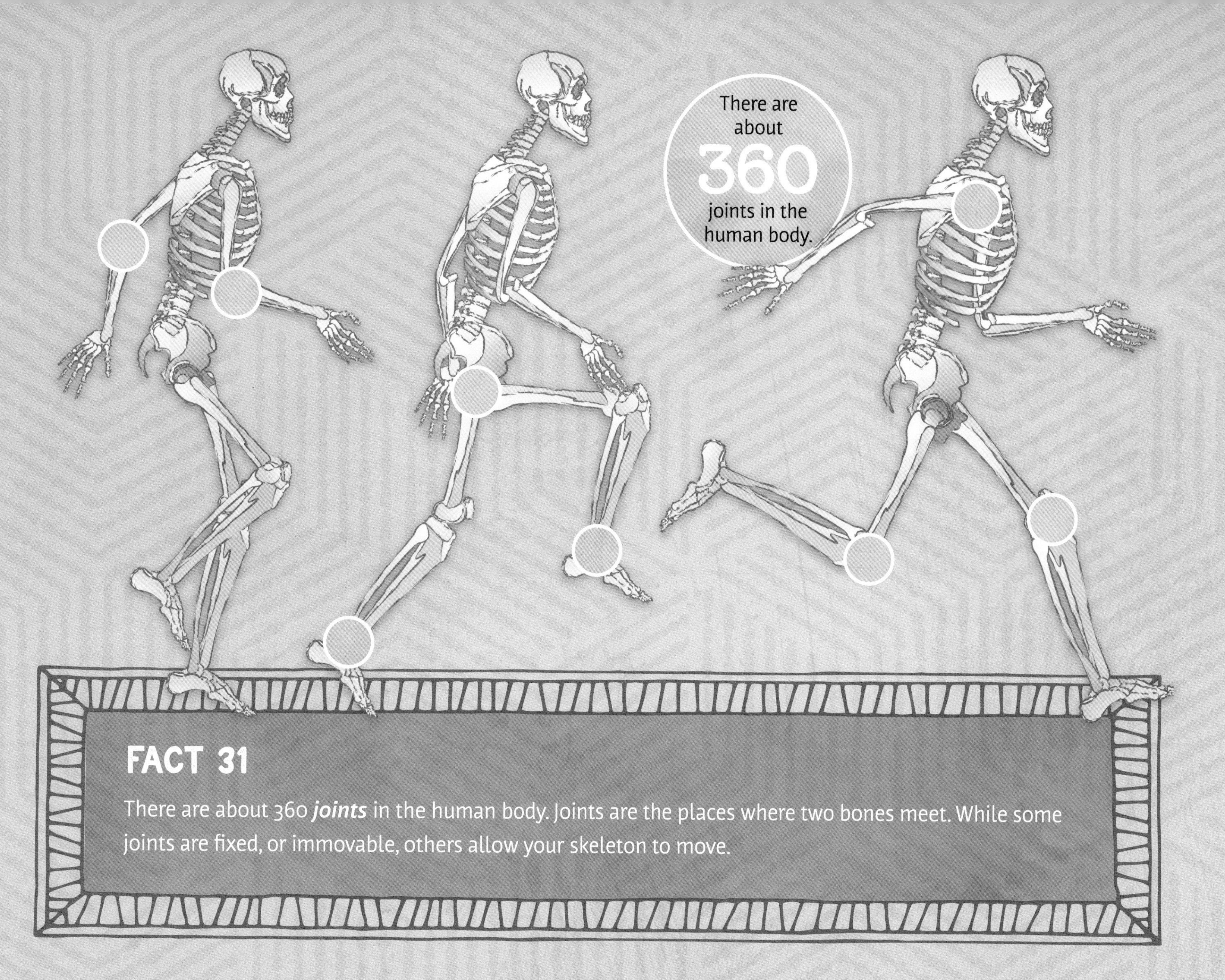

FACT 31

There are about 360 ***joints*** in the human body. Joints are the places where two bones meet. While some joints are fixed, or immovable, others allow your skeleton to move.

FACT 32

The knee is the largest joint in the human body and is also where the two longest bones in your body meet: the femur and the tibia.

FACT 33

Bones have a protective covering of cartilage and fluids at each movable joint to help reduce friction when the bones move past each other. Without this special padding, your bones would rub painfully against each other.

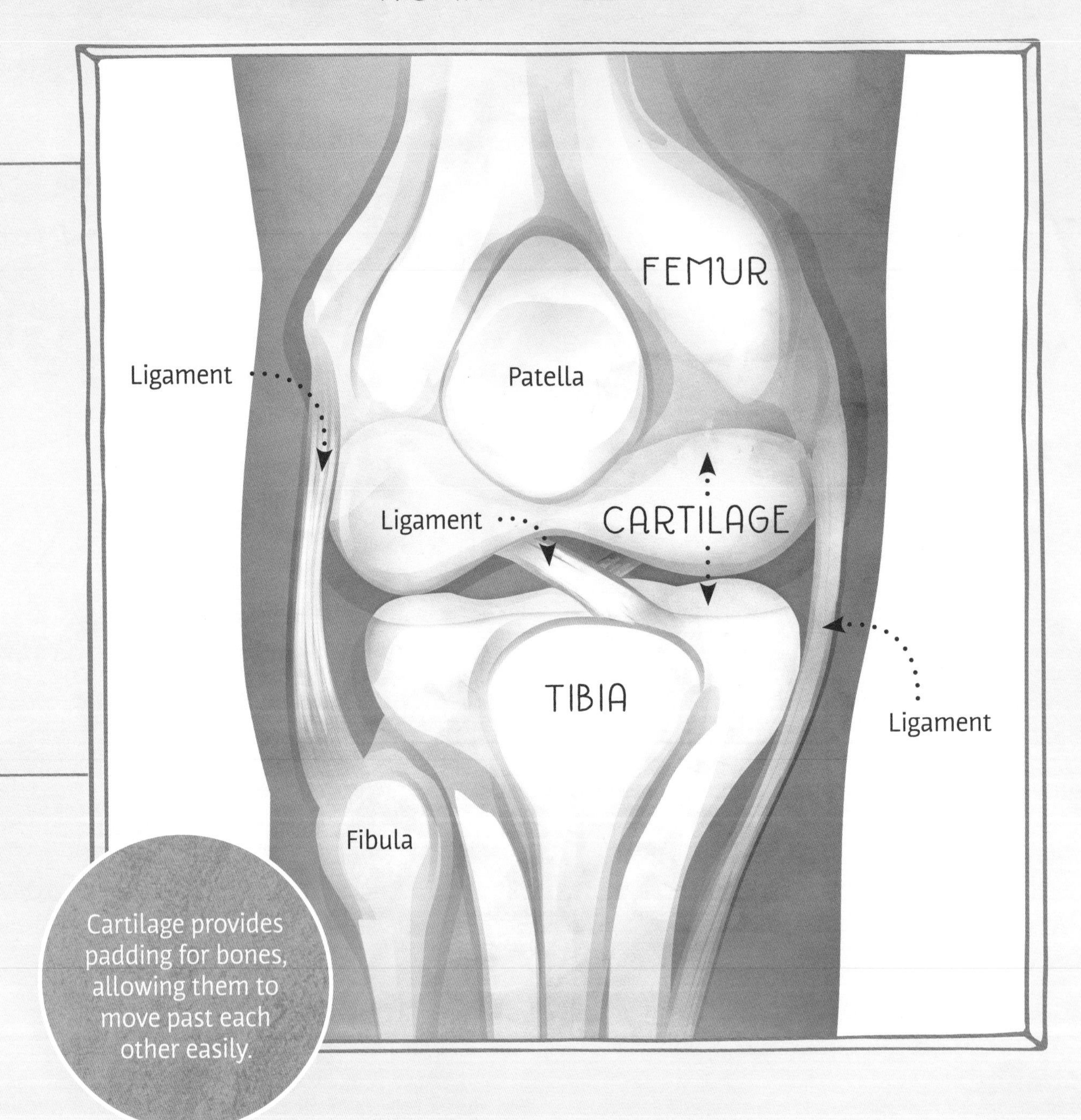

Cartilage provides padding for bones, allowing them to move past each other easily.

FACT 34

Try moving your lower jaw, or ***mandible***, up and down. It's the only bone in your head that moves! It is also the largest and strongest bone in your face.

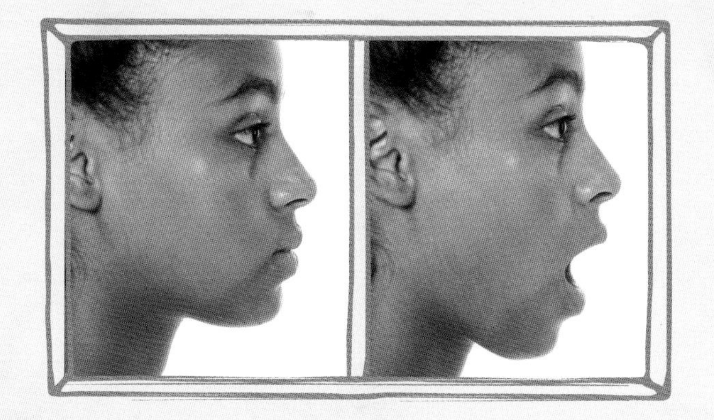

FACT 35

There are six types of movable, or ***synovial*** [suh-NO-vee-uhl], joints:

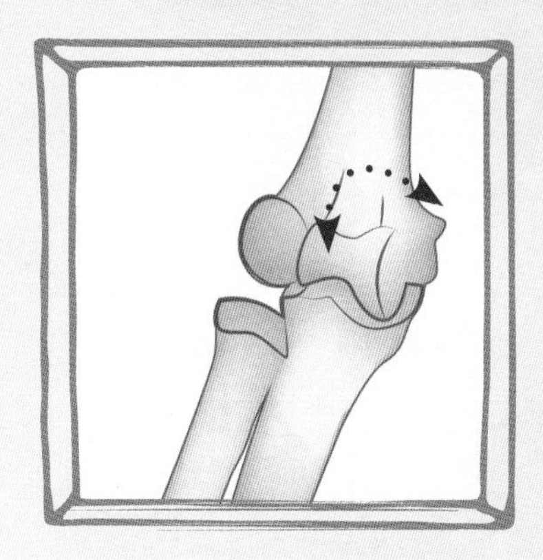

HINGE

knees, elbows, fingers, toes

PIVOT

neck

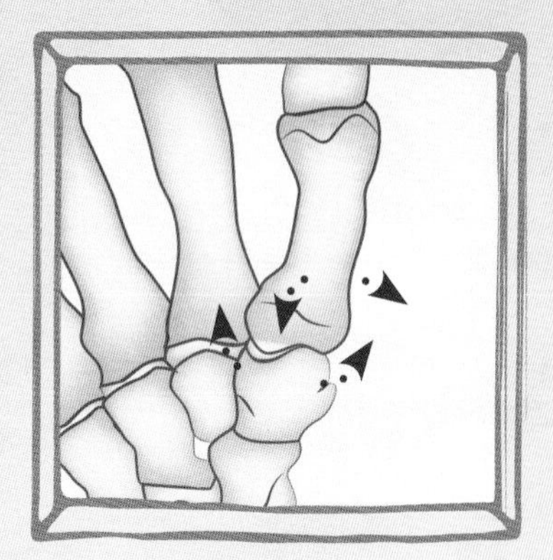

SADDLE

thumbs

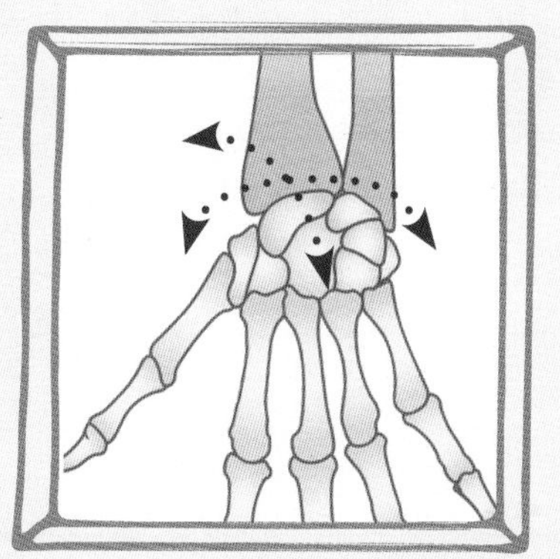

ELLIPSOIDAL/ CONDYLOID

wrists

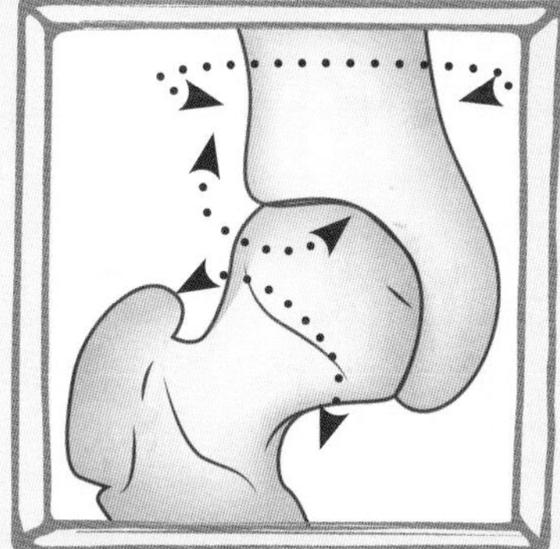

BALL-AND-SOCKET

shoulders, hips

GLIDING/ PLANE

hands, feet, spine

FACT 36

There is only one bone in your body not connected to another bone: the ***hyoid bone***. It's a U-shaped bone found at the base of your tongue.

HYOID BONE
(shown in red)

COMPOSITION

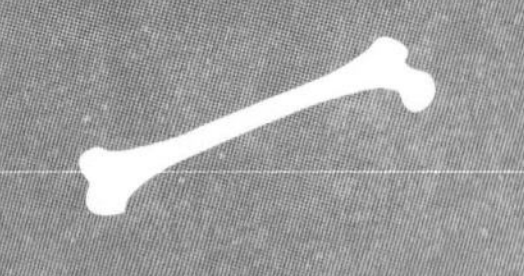

FACT 37

The hard part of bone is made up of an extremely strong, yet flexible, protein called ***collagen***. This is reinforced by the mineral called calcium to add greater resiliency and hardness to the bone's framework.

COLLAGEN MOLECULE STRUCTURE

BONE CONTENT

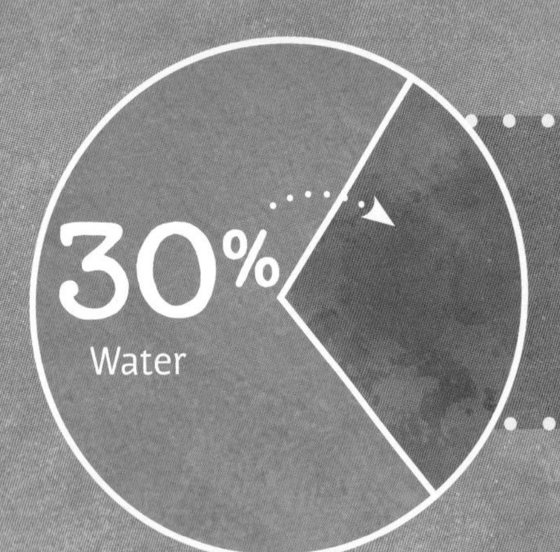

FACT 38

Bones are made up of about 30% water.

FACT 39

Your bones act as storehouses for calcium and phosphorus, which are important minerals for many of your body's functions. When your body needs these minerals, your bones release them into your bloodstream.

RELEASE OF MINERALS INTO BLOODSTREAM

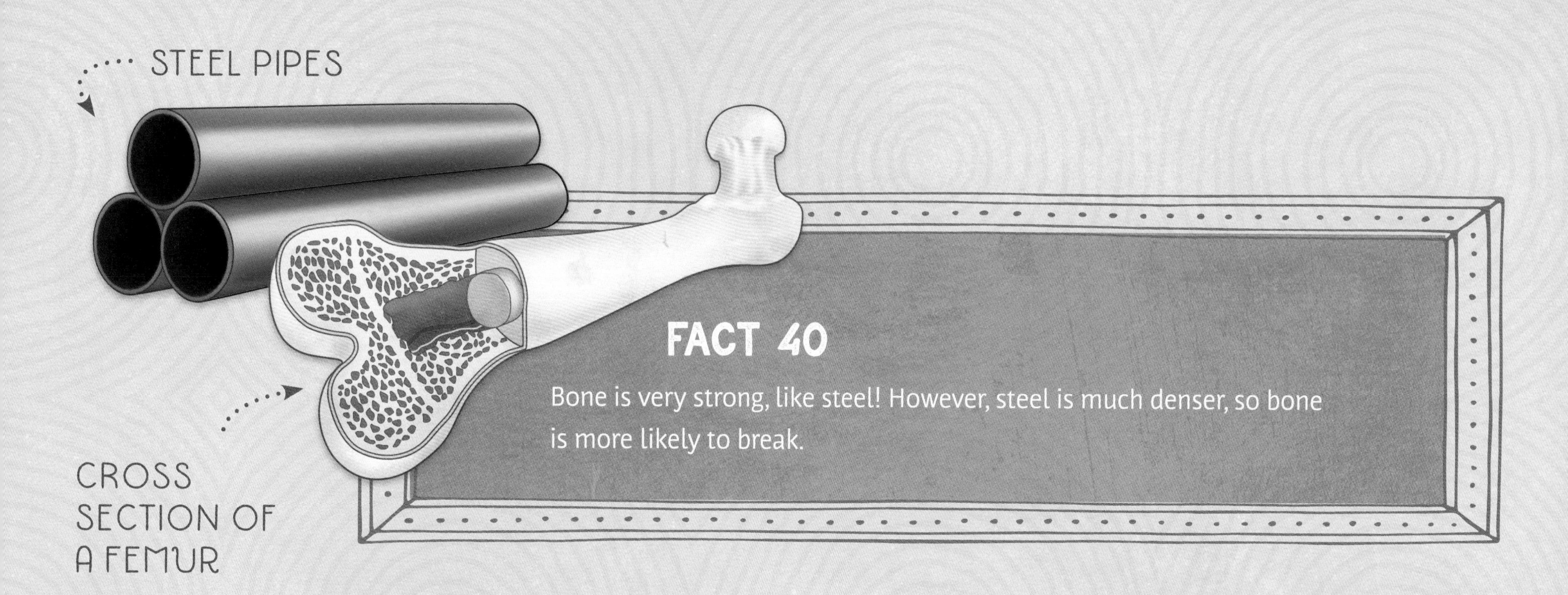

FACT 40

Bone is very strong, like steel! However, steel is much denser, so bone is more likely to break.

FACT 41

Teeth are similar to bones in terms of composition, but they are not actually bones because they can't heal or regenerate once they're broken.

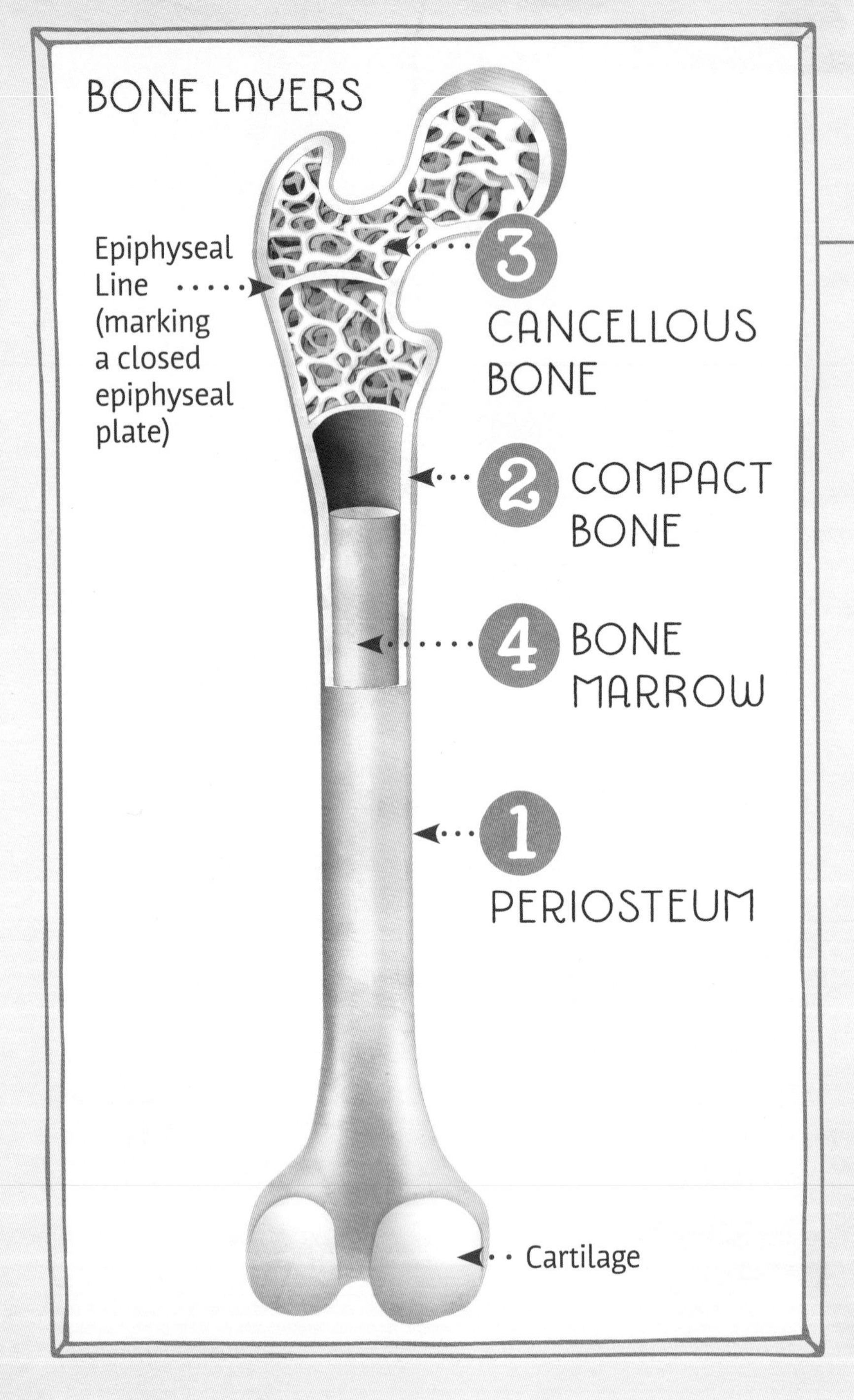

FACT 42

Bones consist of four different layers:

1. The smooth, thin outer layer is a membrane called ***periosteum*** [pare-ee-OSS-tee-um]. It contains the nerves and blood vessels required to nourish the bone and plays an essential role in healing a broken bone.
2. The next layer is ***compact bone***. It is the thickest and the hardest layer, forming the bone's structure and protecting the fragile inner layers.
3. The third layer is the spongy ***cancellous*** [KAN-sell-us] ***bone***. It is light and hard and has a network of interspersed spaces similar to honeycomb, designed to protect the final layer.
4. ***Bone marrow*** is found in the center of the bone. It is soft and jellylike and plays an amazingly crucial role. It is where most of your blood cells are made!

FACT 43

Bone marrow produces more than 200 billion new blood cells every single day.

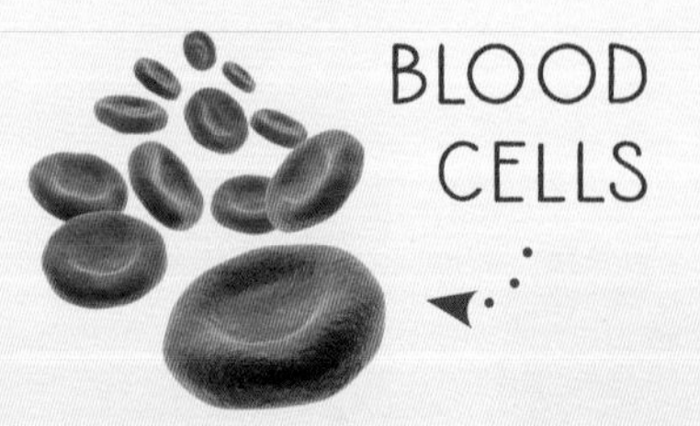

X-RAY OF AN ADULT ARM SHOWING A FRACTURED BONE

FACT 44

There are many different types of bone breaks, also called ***fractures***. An ***X-ray*** can help a doctor determine if a bone is broken and show the type of break it is by taking a picture that goes through your skin down to your bones!

BONE BREAKS

FACT 45

Sometimes a broken bone can be moved back into place by a doctor, and a stiff ***cast*** will be worn to keep the bone in position to allow it to heal straight. Not all broken bones can be placed in casts, however. Broken toes, ribs, or collarbones are examples of bones that may have to heal by themselves. Sometimes a broken toe will be taped to a neighboring toe, while a broken finger may be put in a special soft cast called a ***splint***. Other times surgery may be required for severe breaks, such as a shattered bone. Pins can be placed around the fracture to hold the bone in place while it heals.

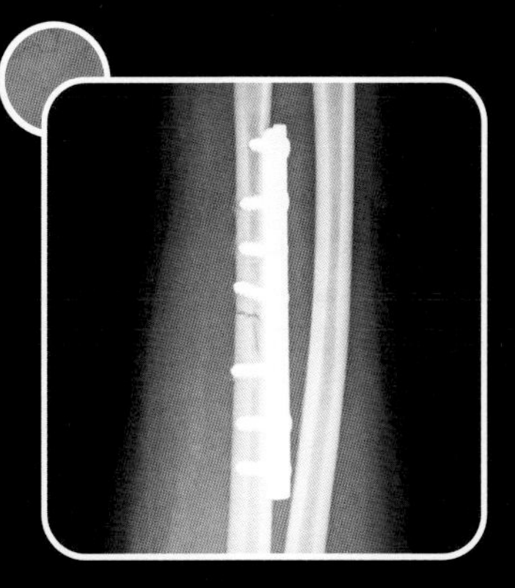

PINS HOLDING A FRACTURED BONE IN PLACE

BROKEN ARM SET IN A CAST

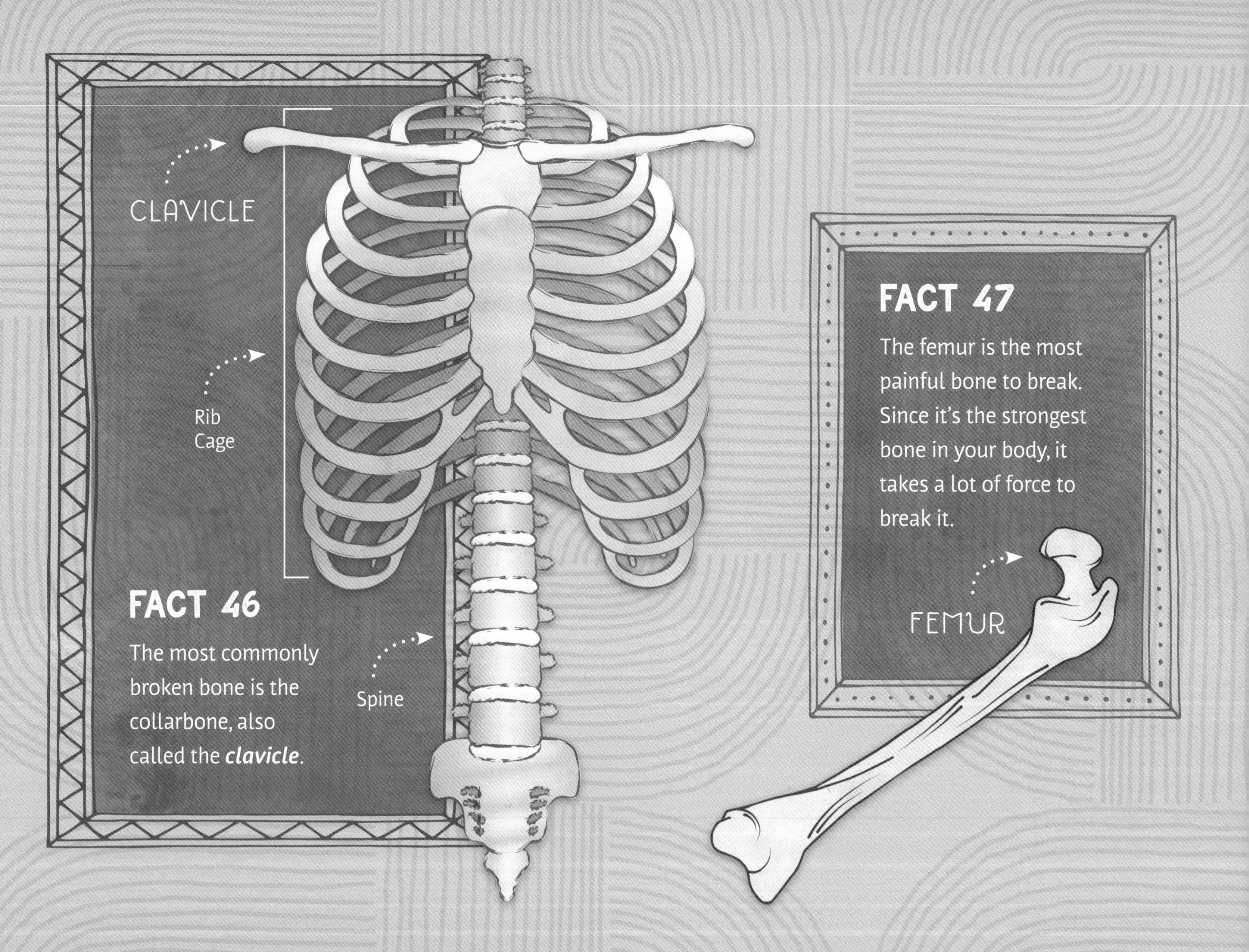

FACT 46

The most commonly broken bone is the collarbone, also called the ***clavicle***.

FACT 47

The femur is the most painful bone to break. Since it's the strongest bone in your body, it takes a lot of force to break it.

BONE HEALING

FACT 48

Because our bones are alive, they heal on their own using a very unique process. At first, blood rushes to the broken area to form a ***clot***. Next, special bone-making cells called ***osteoblasts*** surround the break and begin to form a network of fibers called a ***callus***. The callus is rough and bumpy at first, but it eventually dissolves and smooths out as new bone and blood vessels form to close the gap.

STAGE 1

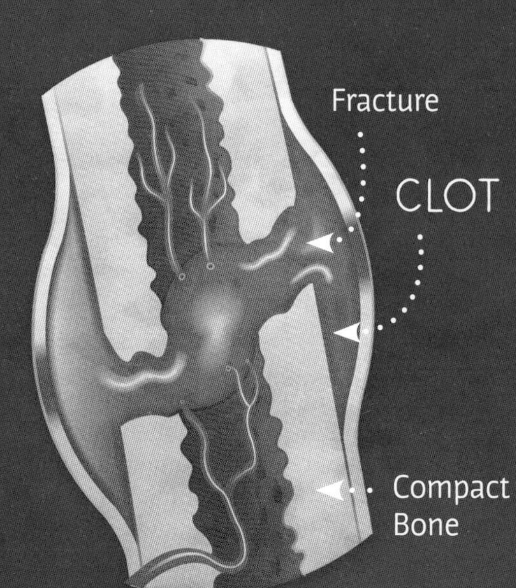

STAGE 2

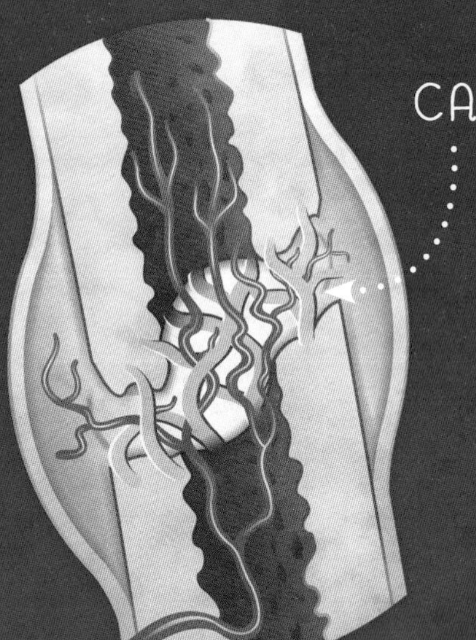

FACT 49

It takes anywhere from 3 to 12 weeks for a broken bone to heal. Children's bones, which are thicker and stronger than adults' bones, typically heal faster. Children also have more periosteum (the outermost vascular layer of a bone), bringing more nutrients and oxygen to the site of the break.

CHILD BONE

ADULT BONE

FACT 50

Scientists have invented ways to implant new bone in patients who have had serious bone diseases or injuries and can no longer grow healthy bone on their own. Sometimes doctors use a technique called a ***bone graft***. This is when doctors use healthy bone from another part of the patient's body or use an artificial bone made in a laboratory to replace the damaged or missing bone.

DENTAL BONE GRAFTING

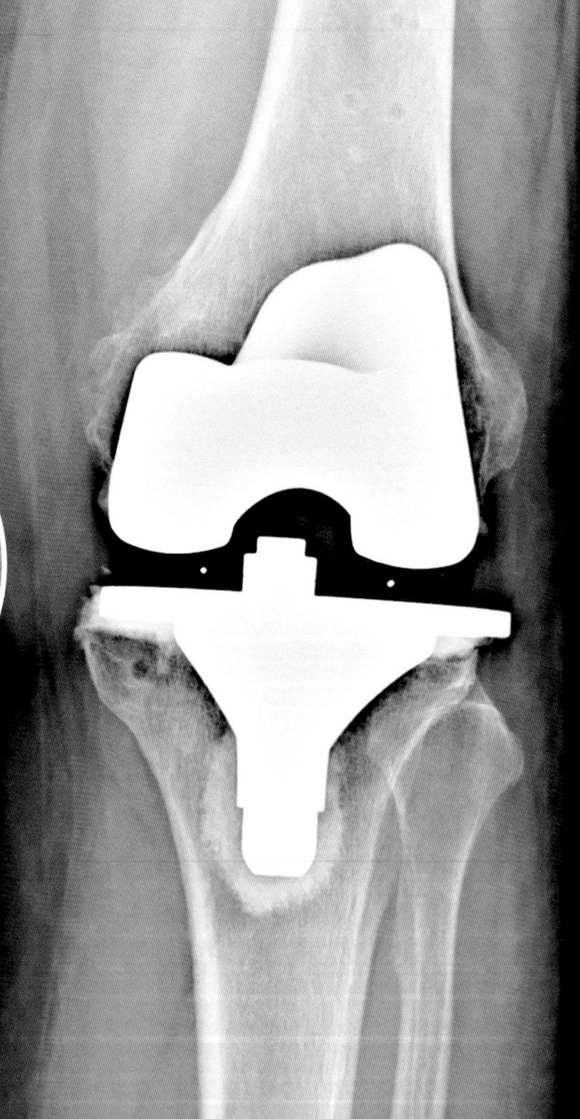

Due to severe bone loss, this knee was surgically reconstructed using an artificial joint and bone grafting.

X-RAY OF AN ADULT KNEE AFTER SURGICAL RECONSTRUCTION

Because of the bones that live and grow within us, we are able to run, wave, talk, jump, stand, and play! Every day, bones help us do wonderful things we wouldn't be able to do without them. They are truly incredible creations.

LEARN MORE ABOUT HEALTH AND THE PHYSICAL BODY WITH THE GOOD AND THE BEAUTIFUL LIBRARY!

GOODANDBEAUTIFUL.COM